CONTENTS

DISASTER

There are different kinds of environmental disasters. Some happen naturally: a volcanic eruption can affect the world's weather for years, for example. A tidal wave can kill hundreds of people and wreck homes and businesses.

Other environmental disasters – the kind this book is about – are caused by humans. We are responsible for two main sorts of disasters. The first are short-term events that happen suddenly. This kind of disaster might be on the television news: a terrible oil spill or a leak from a chemical factory or a nuclear accident.

The second sort of environmental disasters take place over a long period of time. These are less likely to appear on the news because they don't happen as a single event like an oil spill.

▶ *Every day huge amounts of pollution pour into our atmosphere, storing up trouble for future generations.*

DISASTER!
IN THE
ENVIRONMENT

Paul Mason

Belitha Press

Produced by
Roger Coote Publishing
Gissing's Farm, Fressingfield
Suffolk IP21 5SH, UK

First published in the UK in 2002
by Belitha Press A member of **Chrysalis** Books plc
64 Brewery Road, London N7 9NT

Designer: Victoria Webb
Editor: Kate Phelps
Picture Research: Lynda Lines

ISBN 1 84138 411 9

British Library Cataloguing in Publication Data for this book is available from
the British Library.

Printed in Taiwan
10 9 8 7 6 5 4 3 2 1

Acknowledgements
We wish to thank the following individuals and organizations for their help and
assistance and for supplying material in their collections: Bruce Coleman
Collection 14 top (Luiz Claudio Marigo); Corbis 8-9 (Joseph Sohm), 26
(Bettmann), 29 (Joseph Sohm), 30 (Gary Braasch); Corbis Digital Stock 18-19
(Marty Snyderman); Digital Vision front cover, back cover top, 2-3, 4, 5 top, 5
bottom; FLPA 11 top (Mark Newman), 16 (Sunset), 22 top (Minden Pictures);
MPM Images 27; Rex Features 1 (Sipa), 11 bottom (Sipa), 12 bottom (Sipa), 13
(Sipa), 21, 24 (Action Press), 25 top (Sipa); Still Pictures 7 top (Sabine Vielmo), 9
(Klaus Andrews), 14 bottom (John Maier), 15 (Mark Edwards), 17 top (Helmut
Schwarzbach), 17 bottom (Mike Schroder), 18 bottom (Roland Seitre), 19 bottom
(Sonja Iskov), 20 (Gil Moti), 23 (Martin Pruneville), 25 bottom (Mark Edwards),
28 (Julio Etchart); Topham Picturepoint back cover bottom left, back cover bottom
right (AP), 6, 7 bottom (AP), 10, 12 top (AP), 22 bottom (Image Works), 31.

▼ *Thick smog hangs over the city
of Los Angeles in the USA.*

LOOK FOR THE GLOBE

Look for the globe in boxes like this. Here you
will find extra facts, stories and other interesting
information about disasters in the environment.

An oil tanker runs aground, spilling part of its cargo into the sea. Disasters like this have a terrible short-term effect on sea life and birds. Find out more on pages 10–11.

▼ *Loss of the world's forests is an environmental catastrophe that continues to get worse. Find out more on pages 14–15.*

But in some ways these long-term disasters are more serious. They take a long time to sort out because they build up over a long time.

Disasters like this include the effect our industries are having on the world's climate, loss of some of the world's species of animals and the destruction of the world's forests.

🌐 WOLVES IN THE WOODS!

Red kites, birds of prey that were once common in Britain, almost died out a few years ago. Young chicks were brought in from Spain and Sweden, and the number of kites began to grow again. Now some people wonder if other species that have died out could be brought back. Imagine if there were wolves and bears in British woods again!

CHERNOBYL MELTDOWN

On 26 April 1986, one of the world's worst nightmares came true. Ever since the world began to use nuclear energy, people have feared a terrible catastrophe at a nuclear power plant.

Nuclear power stations provide relatively cheap power and, when running properly, do not harm the environment. But if radioactive materials escape from the power station, they cause devastation, poisoning the land and killing people and animals.

At Chernobyl, a nuclear power station in Ukraine, a series of mistakes was made during an experiment in 1986. The mistakes quickly led to three explosions in the heart of the power station. Thirty-one people were killed immediately, but worse followed.

▼ *The reactor at Chernobyl after the disastrous explosion in 1986.*

THREE MILE ISLAND

In 1979, there was an accident at the Three Mile Island nuclear power station in Pennsylvania, USA. Fears of radioactive leaks caused an outcry: the public was faced for the first time with the possibility of a nuclear disaster.

Everyone living within 30 km was evacuated for fear of the radioactive poison that had been released. But the radioactivity rose on the wind and spread over a vast area. Right across Europe and the USSR, people, animals and food crops were affected. The power station is no longer in operation but it is still surrounded by hundreds of kilometres of poisoned land, on which several million people live.

▲ *A cow grazes beside a sign warning that the land around Chernobyl is contaminated by radiation.*

◄ *After the Chernobyl disaster, local people are tested for the presence of radiation, which can cause cancer and other illnesses.*
On 16 December 2000, the reactors at Chernobyl were finally shut down.

THE GLOBAL GREENHOUSE

The Earth's average temperature is slowly getting warmer. This slow warming-up is caused by the greenhouse effect. Warmer weather may sound like a great idea, but it could turn out to be the worst environmental catastrophe of all.

The greenhouse effect is caused by a build-up of certain gases in the Earth's atmosphere. The gases are called greenhouse gases because they keep heat in like the glass of a greenhouse. For thousands of years, they have kept the right amount of heat inside the atmosphere for life on Earth to continue.

 VENUS HEAT TRAP

One example of a greenhouse effect gone crazy is the atmosphere of Venus. Venus has an atmosphere thick with carbon dioxide, a greenhouse gas. It keeps in so much heat that the temperature on the planet's surface is hot enough to melt lead!

▲ Huge queues of traffic on a freeway in the USA. Exhaust fumes from vehicles add to the amount of greenhouse gases in the Earth's atmosphere.

Now human activities – industrial waste gases and exhaust fumes from cars, for example – are increasing the amounts of greenhouse gases. This could be a disaster, as it is slowly causing temperatures to rise. One effect of this could be that the polar ice caps melt, releasing the water they contain. This would lead to large areas of coastal land being flooded.

◄ Researchers in the Arctic take samples of the ice. In recent years the Arctic and Antarctic ice caps have begun to shrink because of rising temperatures.

ALASKAN OIL SPILL

In March 1989, the oil tanker *Exxon Valdez* ran aground in Prince William Sound in Alaska. The hull of the tanker was ripped open, and oil poured out into the clear waters of the Gulf of Alaska. It was the worst-ever American oil spill.

Big oil spills can ruin huge areas of coastline. The creatures that live there are coated in thick oil. Before the flow of oil from the hull of the *Exxon Valdez* stopped, 49 300 000 litres had been released.

▼ The Exxon Valdez *aground in Prince William Sound. Oil spills such as this one are catastrophic for the nearby environment, affecting sea creatures, birds and people.*

EUROPE'S WORST OIL SPILL

In 1978, the oil tanker *Amoco Cadiz* ran aground in the English Channel. Over 305 000 000 litres of oil washed into the sea – six times as much as from the *Exxon Valdez*. Ten years later the oil was still appearing on beaches and having an effect on fishing and tourism.

▼ *A sea otter covered in oil as a result of the* Exxon Valdez *spill.*

Prince William Sound is an important fishing area and has many different kinds of wildlife. Volunteers and other workers rushed to the Alaskan coast to try and help. They rescued animals and birds that had been coated with oil and removed oil from the beaches and rocks by the bucketload.

▼ *Professionals and volunteer workers rushed to Alaska to help try to clean up oil, but the effects of the spill continued for years.*

Despite the huge efforts made after the disaster, the US government found two years later that animals and birds living along the shoreline were still being affected by oil from the spill.

KILLER GAS IN BHOPAL

On 3 December 1984, inside the Union Carbide factory in Bhopal, India, water is somehow mixed into a tank of methyl isocyanate. A huge cloud of poisonous gases begins to pour out of the tank. It is the start of the worst industrial catastrophe ever.

As the people of Bhopal slept, the cloud of killer gas rose up into the air. The wind spread it over a large area. Within hours over 2000 people were dead, and many more – as many as 200 000 – were affected by the disaster. Large numbers of those who survived were blinded; others had problems breathing after their contact with the gas.

▲ *After the Bhopal disaster many of the victims were left with stinging eyes or were blinded by the gas that was released.*

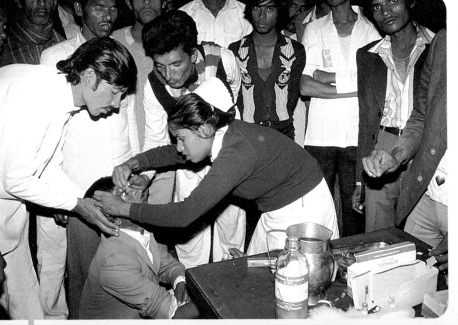

◄ *A nurse helps one of the victims of the gas in an emergency clinic in Bhopal.*

▲ *The Union Carbide factory from which the poisonous gas escaped.*

The effects of the gas leak are still felt in Bhopal. Up to 18 000 people have now died because of illnesses caused by the gas leak. Campaigners say that poison from the factory is still creeping into the local water system, making people sick. And many of the survivors are still waiting for payments to make up for the catastrophe.

 SABOTAGE?

Even nearly 20 years after the Bhopal disaster, no one is sure how it happened. Union Carbide claims it was started deliberately by an unhappy worker. Others claim that the company is simply trying to avoid taking responsibility for the catastrophe.

RAINFOREST CATASTROPHE

In the 1970s, the Brazilian government built the Perimetral Norte (Northern Edge) road. It sliced through the Amazon rainforest home of the Yanomami people. The road was the worst catastrophe in Yanomami history.

▼ *A Yanomami man and two children in their village in the Amazon rainforest.*

▼ *Miners in the Amazon use high-pressure jets of water to wash away soil. This destroys the riverbanks, and the miners bring diseases into the rainforest.*

▶ *Plants that can be used as medicines on sale. The rainforest is full of such plants, some of which have not yet been identified. If the forest is destroyed, these plants may be lost forever.*

The road brought many disasters to Yanomami lands. Diseases such as influenza, measles and tuberculosis arrived. The Yanomami had no resistance to these diseases, and many hundreds of them died. By 1989, about 45 000 gold miners had invaded the land, destroying riverbanks and polluting the water with chemicals.

The Yanomami are not the only people affected by activity in the rainforest. One of the main industries in the forest is logging. Vast areas of trees are cut down each year, which could have a terrible effect on the world's climate. Trees take gases out of the air which would otherwise cause the world's temperature to rise (see page 8). This could have a disastrous effect.

The rainforests also contain undiscovered plants. Some of these could provide new cures for illnesses, but if the forest is cut down they will never be found.

 MINING THREATS

Other Amazon people who are threatened by miners include:
• The Harakmbut in Peru, Ye'cuana in Venezuela and Chiquitanos in Bolivia.
• The Awa Guaja, Uru Eu Wau Wau, Waimiri-Atroari and Parakana in Brazil.

THE RUBBISH MOUNTAIN

Almost everything we buy today comes with some sort of packaging. Chocolate biscuits, bread, shoes, drinks – all come wrapped in paper or plastic. It's becoming almost impossible to get rid of. One day we may be buried under a mountain of our own rubbish!

In the USA alone, homes and businesses produce over 162 million tonnes of waste per year. This figure doesn't include waste from mining, industry or agriculture. Other Western countries produce similar amounts of rubbish: the equivalent of 18 kg of waste per person.

▼ *Mountains of rubbish are tipped into landfill sites like this one every month.*

Our rubbish is usually buried in landfills. These are giant holes in the ground, which are filled with rubbish and then covered over. Landfills cause problems because they leak dangerous chemicals that get into the local water system. Landfills have also caused explosions: gases build up underground and then catch fire. The explosions are like a bomb going off underground.

What to do with our waste is a huge problem. Burning, reusing and recycling waste can all help, but the only real solution is for each of us to produce less rubbish.

▲ *Very few people want to have an incinerator which burns waste like this one in Hamburg, Germany near their home, so building new incinerators is increasingly difficult.*

▼ *Avoiding products that are wrapped in lots of packaging cuts down on waste.*

THE WORLD'S WHALES

From the early 1600s, ships began to set sail with only one aim: to kill as many whales as they could. By the late 1800s, whaling ships had stripped the oceans almost bare of whales.

Whales were hunted for several reasons. Whale oil was used in lamps and provided light for millions of people. Whalebone was used as part of women's clothing. Ambergris, found in the guts of the sperm whale, was used in perfumes.

Whaling ships stayed at sea for up to four years – it could take that long to get a full load of whale oil. The whalers concentrated on one species of whale at a time. Sperm whales, right whales (so called because they were the 'right' whale to catch) and bowhead whales were almost wiped out by the mid-1800s.

◄ *Whaling was a dangerous business for men as well as whales, but in the end the whales suffered far more than the whalers.*

HENDERSON ISLAND

In 1820, a group of whalers landed on Henderson Island in the Pacific. Their ship, the *Essex*, had been sunk by a whale. Within five days the 20 whalers had eaten all the island's grass and bird's eggs they could find, and the birds themselves had left.

▼ A humpback whale swims through the waters of the Pacific Ocean.

▼ Pilot whales being brought ashore after the annual whale hunt in the Faroe Islands.

In the 1900s, numbers of whales increased and whaling started again. Huge ships once more began killing whales for their oil and meat. Blue whales, humpbacks and sperm whales were all pushed close to extinction as over 30 000 whales a year were killed. The terrible loss of whales only stopped in 1982, when whaling for money was banned.

THE SHRINKING SEA

One of the most amazing environmental disasters is the story of the Aral Sea, which borders Kazakhstan and Uzbekistan. Between 1960 and the mid-1990s, the sea shrank to half its original size.

▲ *Deserted ships stranded tens of kilometres from the shore of the Aral Sea.*

Two rivers feed the Aral Sea. Water is taken from these rivers before they reach the Aral to water crops and to supply towns and villages. The amount of water that reaches the sea is now far less than it was before.

The sea is saltier and more polluted as a result of having less water. This means there are fewer fish in the sea each year, which has had a terrible effect on the local fishing industry.

Towns that once sat on the edge of the Aral Sea have been left high and dry. The town of Muynak was once a port but is now over 70 km from the seashore. The sea's shrinkage has been a disaster for the surrounding area. Chemicals from the exposed seabed are blown across the land by dust storms, causing pollution.

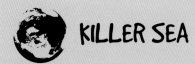

KILLER SEA

As the Aral Sea becomes saltier many of the animals that live there are dying out. They cannot survive in the salty water. Among the animals affected is a unique species of dolphin.

▼ *The Aral Sea has lost over two-thirds of its water since 1960; areas that once made up the sea bottom have become deserts.*

GREAT SURVIVORS

It can seem that we are surrounded by environmental disasters. Oil spills, nuclear leaks, threats to wildlife and pollution all appear in the news regularly. But there are some success stories too – places or animals that have survived the threats.

Two rivers run either side of Manhattan island, the heart of New York City, USA. The Hudson River and the East River were both once so dirty that almost all fish died out. They could not survive in the water, which carried pollution from upstream as well as from the city's industries. Both rivers were cleaned up during the 1980s and 1990s and are now clean enough for fish to live in once more.

▲ A grey wolf, which at one time had almost died out in Europe. Wolves are now making a comeback, with growing populations in Poland, the Balkans and Romania.

▼ New York's East River, which at one time was heavily polluted.

▶ *This oil spill in December 1999 on France's north-western coast was a potential disaster. A mixture of good luck and hard work meant that local wildlife was not affected as badly as had been feared.*

Five species of sea turtle lay their eggs on Brazilian beaches. In the past, almost all those coming ashore were killed. In 1980, Project Tamar began to persuade local people not to kill the turtles. Now the locals protect the turtles, and each season around 300 000 baby turtles are released into the sea.

 ## SIBERIAN OIL SPILL

In 1994, an oil pipeline carrying oil across northern Siberia split. Oil began to pump out into the local environment. The Russian government and international agencies immediately began work to clean up the spill. Prompt action and good luck with the weather meant that they were able to prevent an environmental catastrophe.

In 1997, a United Nations Conference on climate change met in Kyoto, Japan, to try and limit the amount of greenhouse gases released into the environment. The countries agreed to reduce the amount of greenhouse gases they release to 6–8 per cent of 1990 amounts. Even though the USA has since changed its mind about making these reductions, the Kyoto Agreement will still help slow down the greenhouse effect.

ENVIRONMENTAL RESCUE

All round the world people are fighting to protect the environment. Some are volunteers who give up their time when there has been a catastrophe.

Others belong to organizations that try to find long-term solutions to environmental problems like the greenhouse effect or loss of the world's rainforests.

DIAN FOSSEY

The naturalist Dian Fossey worked for years with mountain gorillas, which are an endangered species. Fossey has worked to make people aware of their situation. Her story was even made into a film called *Gorillas in the Mist*.

▼ *Members of the environmental organization Greenpeace occupied this old oil platform in 2000 to stop it being sunk and left at the bottom of the sea.*

The musician Sting alongside Yanomami campaigners trying to preserve the rainforest.

Imagine hearing that there had been a huge oil spill near your home. Would you go and help with the clean-up operation? After disasters like the *Exxon Valdez* oil spill, thousands of people volunteer to help. They give their time in an effort to help the environment. Sometimes famous people try to help too. The musician Sting, for example, campaigns to stop the destruction of the Amazon rainforest. The actress Pamela Anderson campaigns for animal rights.

Environmental campaigners object to US President George W Bush, who they accuse of putting money ahead of the environment.

Large organizations work to protect the environment from disaster in two ways. They help when a catastrophe such as an oil spill happens. They also campaign for long-term changes that will prevent disasters from happening in the future. This might include trying to stop animal habitats being destroyed, trying to encourage people to create less waste or campaigning to reduce greenhouse gases.

DISASTER FACTS

The biggest-ever oil spill was at the Ixtoc I *oil well in the Gulf of Mexico, in 1979.*

SECRET DISCOVERY

Whalers and cod fishermen from the Basque country in Spain may have discovered America before Columbus! Many Basques say that their fishermen found the new continent in the 1300s but kept it secret because they didn't want other people to know about the rich fishing grounds offshore.

SAVED BY PETROL

The whaling industry shrank after 1859 when oil (petroleum) was discovered. Whale oil was no longer in demand for lighting, which probably saved many species of whale from extinction.

CAR POLLUTION

The discovery of petroleum in 1859 brought new environmental problems, especially the increase in greenhouse gases and the effect on animal habitats of road building.

BIGGEST SPILL

The largest recorded oil spill was the blow out in 1979 of the *Ixtoc I* oil well in the Gulf of Mexico. It released about 630 000 000 litres of oil into the sea and seemed likely to cause a terrible catastrophe. In the end, the oil was blown out to sea by the weather and did not affect the coastline.

Noise Nuisance
One of the fastest-growing types of pollution is invisible: noise pollution. Background noise increased by one decibel per year in the 1980s. In a typical Western home today, average background noise is about 45 decibels. If things continue at this rate, by 2085 it will be so noisy that it will physically hurt!

▼ *One of the first oilfields in US history, Oil Creek in Pennsylvania in 1865. Within a hundred years the world would be unable to function without supplies of oil.*

DISASTER WORDS

Atmosphere The group of gases that surrounds the Earth, or any other planet, is called its atmosphere.

Climate The usual weather conditions – including sunshine, rain and temperature – in a particular place are called its climate.

Decibel A measurement of the loudness of sound.

Endangered Threatened with extinction.

Environment The surroundings in which people live. Pollution can damage the environment.

Evacuated When people are removed from an area for their own safety we say they have been evacuated. In Britain during the Second World War, children were evacuated from some towns to keep them safe from air raids, for example.

Extinction When a species completely disappears it is said to be extinct. Some species die out naturally, but today many are wiped out by human activity.

Fishing grounds Areas in which there are a particularly large number of fish. One of the most famous fishing grounds is the Grand Banks, which are off the eastern coast of North America.

Greenhouse gases A group of gases that make up part of the Earth's atmosphere. Together, they trap heat inside the atmosphere, making it warm enough for life to survive. But if there are too many greenhouse gases the Earth's climate heats up.

◄ *Litter washed up on the Caribbean coast of Venezuela. Rubbish spreads from one country to another on ocean currents.*

One way to help the environment is to recycle glass and plastics like this woman in Santa Monica, California, rather than simply throwing them away.

Habitat The special surroundings in which something lives are called its habitat. For example, some animals live in a woodland habitat, while others can only survive in a wetland.

Influenza A disease that causes high temperature, shaking, aching limbs and sometimes death. Influenza is easily passed from one person to another.

Measles A disease that causes sickness and red spots to appear on the sufferer's skin. Measles is easily passed from one person to another.

Packaging The wrappers or containers in which goods are supplied are called packaging.

Radioactivity Radioactivity is caused by atoms breaking down and is one of the products of making nuclear energy. It is invisible, but even quite tiny amounts are dangerous to living creatures.

Recycling Recycling involves the reuse of materials that have already been made into something. For example, old newspapers can be recycled to make toilet paper, which saves having to cut down trees (from which paper is made).

Resistance A person is resistant to a disease when his or her body can fight it off.

Run aground A boat runs aground when it hits the seabed and crashes.

Tidal wave A huge wave caused by an earthquake.

Tuberculosis A lung disease which eventually kills its victims unless they are treated. Tuberculosis spreads quickly from one person to the next.

USSR Also called the Soviet Union. A very large, powerful country in Eastern Europe which in 1991 split into lots of smaller countries, including Russia and Ukraine.

DISASTER PROJECTS

BE AN ENVIRO-REPORTER

Pick one of the environmental catastrophes from this book and try to find out more about it. You may find information in other books and encyclopedias and on the Internet. Now try to write a report for a newspaper, saying what has happened, what the effect might be and what is being done.

Alternatively, you could try to imagine yourself as one of the characters in a story: a whaler from the *Essex* on Henderson Island, for example (see page 19) or a worker trying to help with the clean-up after the *Exxon Valdez* oil spill (page 10). What would it have been like to be there? Write a diary account of what you saw and felt, trying to imagine what the weather was like, what animals you saw and how the people around you felt.

▼ *Workers clean up a cormorant that has been covered in oil from the* Exxon Valdez *tanker.*

The more things you reuse and recycle, the less rubbish will be dumped in landfill sites like this one.

REDUCE THE RUBBISH MOUNTAIN

Make a list of all the packaging that's thrown away in your home each week. This could include plastic shopping bags, wrappers, boxes and envelopes. Does all of it need to be thrown away or could some of it – the envelopes or shopping bags, for example – be used again? Did you need all of it in the first place?

CAMPAIGN FOR THE ENVIRONMENT

There are lots of organizations campaigning for the environment. Some of them have branches for young people, which you can join to find out more information and help with environmental projects:

The World Wildlife Fund:
www.wwf.org.uk

Greenpeace: www.greenpeace.org.uk

Surfers Against Sewage:
http://sas.sw.concept2100.co.uk

The National Trust:
www.nationaltrust.org.uk

The National Trust also has a coastline section for children:
http://coastline.nationaltrust.org.uk/kids/index

Try to reduce the amount of packaging you use. You will be able to tell if you're successful because your weekly list will get shorter! Make sure any packaging that you do throw away is recycled.

INDEX